Synthesis as the Principle Factor of the Development of the Universe

By

Vladimir Groh

The most incomprehensible thing about the world is that it is comprehensible.
Albert Einstein

You do not need to leave your room. Remain sitting at your table and listen. Do not even listen, simply wait, be quiet, still and solitary. The world will freely offer itself to you to be unmasked, it has no choice.
Franz Kafka

And ye shall know the truth, and the truth shall make you free.
John, 8:32

Table of Contents

Introduction

Advocates of the Big Bang theory claim that all matter in the Universe up to the time of Big Bang had been concentrated in an infinitesimal volume — in the so called singularity or, in other words, in a point. Further, for some unclear reasons, an explosion took place resulting into the Universe expansion. What we do believe, though, in the beginning was the initial field. And it was a self-developing system where movement had been present, it was not a static, isotropic, and mass-less thing. As a result, dense areas and loose areas begun to evolve in the field. It was one of these dense areas, where the fusion of new objects in the Universe begun to take place (this could very well be the super strings). The fusion is followed by the energy emission and expansion. This looks like an explosion, except that the products of the explosion as it is happening scatter off at a decreasing separation rate, and, in case of an ongoing growth of the fusion zone, this rate will keep increasing because of the energy inflow. This discovery was done by Hubble — the farther the object is from the observer the faster it moves away. This means that the ongoing fusion creates material world, energy is emitted thus resulting into the expansion of the Universe. Thus we suggest that all force interactions of the material world shall be examined in the media evolved as a result of fusion from the initial field. We suggest referring to this medium as to "ether". And this is why we need analyze the ether itself in the first place and its properties upon interactions. When looking at many phenomena and processes that today do not seem to be reasonably founded or cogent, this order of study helps to see them from a new perspective.

1. Levels and Properties of Ether

Let us get clear with the terms. Ether consists of material and non-material levels. Ether level is a set of fields created by quanta with similar physical properties; at that, the ether itself is a superfield.

<u>Material level</u> is a combination of elementary particles — electrons, protons, neutrons, chemical elements, molecules, and macro bodies

<u>Non-material level</u> of the ether consists of zero-mass particles that have, by analogy with fundamental particles, a complex structure and interact in creating their photon fields.

The fusion in Ether is schematically shown in Figure 1.

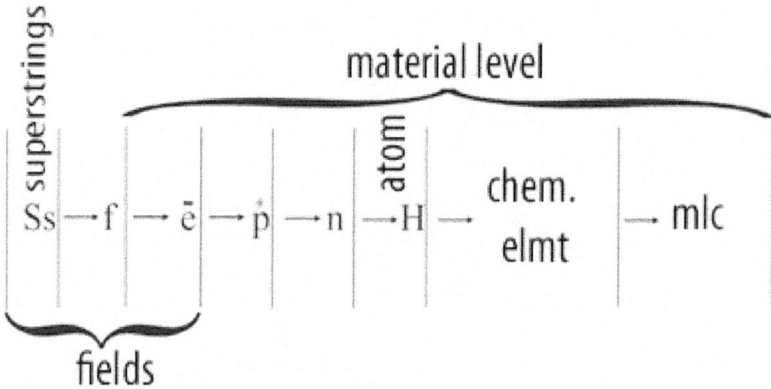

Figure 1. Fusion in Ether

All micro and macro bodies are emerged into the ether including the man itself as an observer (which

4

also has effect on the processes in this super-field). By assuming the existence of the fields we can explain different interactions including relativistic effects. Let us adduce the following proofs of the existence of Ether:

1. Isotropy of radio-frequency radiation from a point source.

2. The deviation of a light ray from a star near the Sun can be explained by the differences in density of all fields created by the Sun at different distances from the Sun, i. e. a regular refraction of ray occurs in environments with different density.

3. Mass defect upon nuclear reactions is a part of the mass that has fitted into another non-material level (the field) and has been realized in a form of a wave in such field. Mass defect can be to the both sides, so there is no need in Higg's bosons to explain how the mass emerges from non-material field:

4. $\Delta E = \Delta m\ c^2 \rightarrow \Delta m = \Delta E/c^2$. The mass change sign (+ or -) depends on whether the energy is emitted or absorbed.

5. Displacement   currents   evolving   in electromagnetic radiation can be explained assuming the existence of the electron field.

6. The velocity of electric current distribution is equal to the velocity of electromagnetic wave propagation within the electron field ($3 \cdot 10^8$ m/sec) rather than the rate of motion of electrons in metal that equals to 0.5 m/sec.

7. Continuous growth of the masses of all space objects can be explained by the continuous matter fusion from the non-material field.

8. Availability of these fields ensures that empty space curvature can be set aside to describe the space bodies interaction by the same formulas of dynamics but in the deformed fields. Field deformation means a local change in its density.

2. Synthesis as the Factor of the Development of the Universe

Scientific evidence gained over the centuries suggests that the matter around us consists of molecules, atoms, elementary particles, and various fields. Obviously, more complex objects of the material world consist of more simple objects. It is proven by experiments held on molecule break-up into atoms - and then experiments in atom break-up into neutrons, protons, and plus quanta of energy. Neutrons break up into protons and electrons, and also quanta of energy. In the experiments using hadron collider, proton was broken up to electron, positron, and γ quantum. (We cover this further in a separate chapter.) The electron consists of smaller particles — quanta of photon field that can be referenced to as "non-material" particles. Why do we call them "non-material"? For the field, which consists of photons, it is considered that they have near-zero kinetic energy is near zero. Accordingly, it does not seem possible to determine photon mass from zero energy. Moreover, current reference guides into physics claim that photons have no rest mass. Such magnitudes could hardly be measured with any instrument but this does not mean that there is no such parameter at all.

It's generally accepted that the fusion is a process where a more complex matter develops from a simple matter. If we assume fusion in the medium in question, then it is obvious that it could be developed from quanta of photon field and towards creation of chemical elements. In fact, there are not enough field quanta in boundary-less photon field to infinitely maintain fusion. Additionally, field is a media which provides for local energy accumulation, which means that the density of the field could be reached at which the fusion begins, see Figure 2.

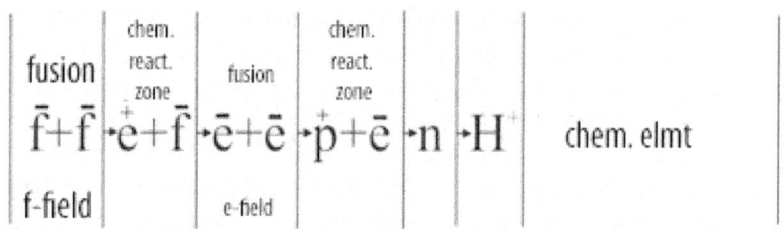

Figure 2. Particles and field fusion

Fusion in the photon field media starts from positron fusion from two photons. Because positron is very active, it takes negatively charged photon almost immediately and turns into electron. Electron and photon with the same charge do not get into reaction with each other resulting into an increase in electrons and a decrease in photons that could easily link with photons. Electron interaction with the photon flux which accelerates toward to the fusion center restrains its further propagation. This shall

8

result into a pressure increase in the electron field that increases toward to the center of electron fusion. Continuous electron fusion leads to an energy increase of the electron field and creates the conditions for proton fusion initiation from electron field. Obviously, there is a zone where electrons and protons interact with creation of neutrons. It can be assumed that the neutron field shall exist in accordance to a pattern in Figure 2. However, we have not shown this field because it could be find only near neutron stars. It is still unclear what could be created as a result of fusion in neutron field.

All we have to do is one further step, assume that a photon field denser than γ-photon field exists. Obviously, it shall be an endless source for photon fusion. The assumption of the presence of "denser" fields allows to explain the model of the Universe's beginning and its dynamic development.

The electrons that could not produce enough energy to generate protons and neutrons are fit into electron field. It must be stressed that electron energy in their fusion zone is much higher then the energy of the electron that create electron field. The energy must be enough to generate protons as well as neutrons. Particle connectivity energy in the field tends to zero, and the energy of the same particle in atom is significantly higher as a result as the dynamic interactions with other particles. The interaction rates between field particles are inversely proportional to their masses. Meaning that the particles in denser fields have relatively smaller weight but a higher interaction rate, and, accordingly, they have high specific energy. On the basis of these assumptions it could be said that fusion in fields can continue indefinitely.

Let us consider the conditions that contribute to the formation of a focal point at the fusion initial stage (singularity). If we study the propagation of electromagnetic disturbance in the ether medium rather than in the vacuum then the magnetic field, which works furthermore as a damper limiting the rate of light propagation, can create the conditions for the local energy accumulations. These areas combined with the potential internal pressure can be exposed to electromagnetic disturbances from several directions both from different sources and from a single source by altering the disturbances directions with the possible manifestation of resonance effects. In the real life, the quantities of the photon field can receive the disturbance pulses at different angles, of different intensity, and so on. Apparently, some of the quantities can twist under the influence of the disturbance vectors to accumulate even more energy. This creates the conditions for a focal point at the fusion initial stage.

In our belief, any particles, and not only the material particles are able to be synthesized upon certain conditions, i. e. they can develop into new stable objects with observed parameters, with fewer quantities, and with the emission of some quantity of energy into another level (mass defect).

It should be noted that as soon as the process of the local fusion related to the mass defect and the reduction of the source capacity of components ends, a step-like reduction of the potential pressure shall occur in this local capacity. This pressure reduction in the specified quantity will be followed by the slowed down fusion until next increase in the value of such pressure to the "threshold" value specific to the initial fusion behavior. So the fusion proceeds in a constant but discrete way.

Photon field creates fluxes that are heading to the fusion area and are moving in these directions with a variable acceleration.

In accordance with Bernoulli's law, the fact that the pressure of the flowing media is reduced back by quadratic dependence at an increased flow rate is correct to describe the dependencies at material media flow and the same is also true of the photon field. It should be noted the fact that as the photon flux approaches nearer the fusion center, the total area of the flux is reduced by quadratic dependence, and this shall logically lead to a significant increase in pressure by quadratic dependence. Therefore, the slight decrease in pressure of the flowing medium caused by the increase in the flow rate by quadratic dependence is excessively compensated by the increase in pressure by quadratic dependence which is due to the change in the geometrical size of the flux. In view of the foregoing, the density of the photon flux significantly increases as it approaches the fusion area; this will lead to the maintenance of media parameters required for fusion.

Conclusion: We have studied the fusion, which evolves and progresses in the medium (in the ether), and have proved that:

1) There are areas where the conditions for the point fusion initialization evolve.

2) The mass defect evolves in the fusion; that is when some of the mass of the initial matter transforms into emission of a denser level with higher energy.

3) There are the conditions when the non-material photons receive the mass and energy comparable with the material particles and create the matter. [1]

4) The accelerating photon field with a vector to the center of the fusion is commonly referred to as gravity.

5) Gravity defines masses for all created material particles.

6) Fusion provides for constant matter growth with generation of space objects and for their change.

7) The medium fusion, where, at least, 3 levels of the ether are affected, deforms it, thus creating the interaction laws of the material world.

It is believed, that the Sun is a thermonuclear reactor where the reactions of helium hydrogen fusion, lithium helium fusion, etc. up to the iron fusion are present; then the fusion is accompanied not by the energy emission but rather by its absorption from the environment. If we calculate on the basis reasoning from this model to determine the time all the hydrogen (equal to the Sun mass) needs to transmit to iron, then the Sun would exist only for 100 million years. Where does the energy, which makes the Sun emit photons, various energies, elementary particles, gases, and larger chemical elements for 4.5 billion years, come from? It seems that the source of such energy is the fusion at a finer level when photons are synthesized into electrons, protons, neutrons, and only then into hydrogen, etc. following the common scheme.

It is generally known, that the Sun produces photons and yet there is a question: what is the source material for the photons? As we have suggested in *Developmental Universality and Unity of the Universe* [1], γ-photons are the source for the photons. It is known, that the highest energy, in accordance with the Planck's law, can be transported by the particles with a higher oscillation frequency. In addition, they are more energetically dense and they are, as we assume, "building blocks" for photons. And they, namely γ-photons, determine the main energy emitted by the Sun. This can be proven by the fact that energy of particle is determined by using its oscillation frequency: $E = h\,v$, and frequency of γ radiation is 10^{10} higher than the photon oscillation frequency, i. e. energy of γ particle is 10^{10} higher than the energy of the light photon.

Of all fields that we have taken for reviewing herein, γ-photon field (superstrings, Ss) is seen as more energetically dense field because it has the higher own oscillation and serves as building blocks for photons as previously mentioned. Photon fusion from γ-photons results in the increased volume and "mass" of the photon field. It causes such conditions that initiate photon fusion: e^+ – positrons, e^- – electrons and formation of the electron field (see Figure 2.) Electrons are fused into protons, p^+. Protons and electrons produce neutrons. The bond energy herein drops because of the pressure drop in the neutron field resulting in transformation of neutron to hydrogen through expanding itself and reducing its energy followed by the events as per the known scheme.

The process of fusion described above occurs also within the planets such as the Earth. They also have their own electronic fields where hydrogen and other chemical elements, including oxygen, are also generated; oxygen and hydrogen upon fusion produce water. This means that water on the Earth rather than being created by the Earth itself is brought in from outside. This can be indirectly proven by the fact that two thirds of the Earth surface are covered with the ocean. This also confirms existence of the authentic fresh-water seas under the Earth surface (from Libya in Africa and to an entire continent, such as Australia). The evidence that matter was created by the Earth is the continuous volcanic eruption; this can be explained by energy increment and emergence of new mass. The reef wrinkles are 250–300 million years old while the Earth itself is 4.5 billion years. But who had brought them to the Earth? Besides, the Earth's diameter and mass keep growing and that does not well correlated with the quantity of space dust-fall.

The diagram of particle and chemical elements emergency on the Sun and on the Earth is shown in Figure 3.

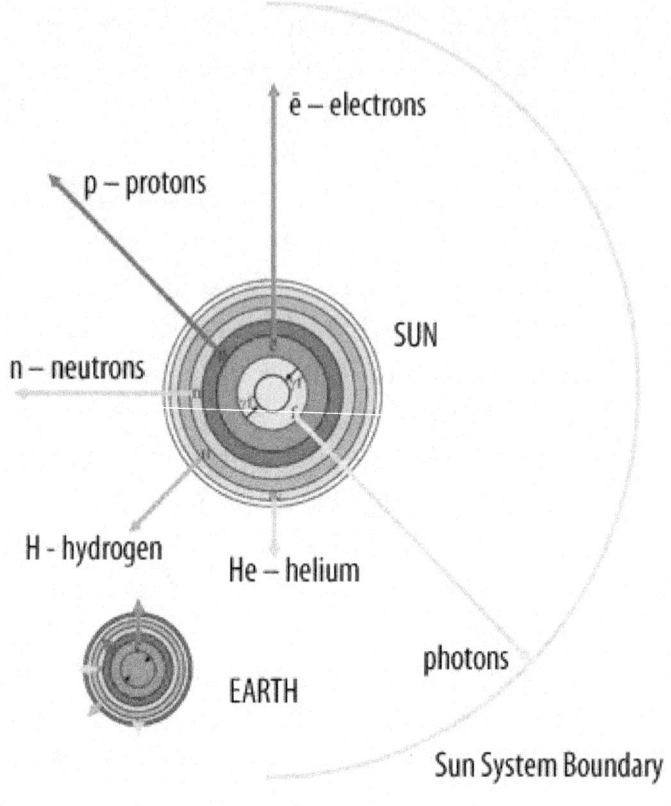

- □ - zone of fusion of photons from γ-photons
- ■ - zone of electron generation
- ■ - zone of fusion of protons from electrons
- □ - zone of neutron generation
- ■ - zone of hydrogen generation
- □ - zone of helium fusion

Figure 3. The diagram of particle and chemical elements emergency on the Sun and on the Earth

3. Fields

The outlined process of field creation is very much in line with Dirac's theory. This theory, which was proposed by Dirac in 1928, integrates quantum mechanics and Special Theory of Relativity. Dirac has found an equation that defines wave function of the electron in relativistic quantum mechanics. In solving this equation, he had faced an unexpected fact. Apparently, along with the Dirac's solutions that well comply with the electron with fully positive energy, there are solutions with fully negative energy. Both solutions are different roots of the same equation. In physics, negative energies are not allowed: when a particle with negative energy collides another particle and transfers part of its energy to it, in this case, it shall gain even more negative energy as an absolute value. Accordingly, it could get as much energies as it takes in an absolute value thus endlessly working on the other bodies (particles) and gaining more velocity of light in vacuum. To resolve this conflict Dirac has proposed a bold hypothesis. He assumed, that all states with negative energies present in vacuum (unless there are no particles with positive energy) are already are full packed with electrons. That is, there is an electronic field with negative energy. Wave in such field will propagate almost immediately and without losses.

Such assumptions is well in line with the effect of entangled twin (created in the same process) electrons or photons — when one of them, by some reason, changes, another one changes accordingly wherever it could be in the Universe. And modern tests give a strong evidence of this effect. In 2008, a Switzer research team from University of Geneva had been able to carry two fluxes of entangled photons 18 km away from each other. This enabled the measurements of travel speed of the particles within the distance to an accuracy that had never been reached before. As a result, it was discovered that even if there is some hidden interaction between entangled photons then the velocity of its propagation shall be at least 10^5 times higher the velocity of light in vacuum. The time delays below this velocity would be observed. Thus, the existence of the Dirac's electron field can be indirectly proven.

All fields are intertwined without precise boundaries and a new field is created within the intertwined fields. For example, proton interacts with electronic field which leads to neutron creation. This means that additional energy is needed to create neutron and electron energy in hydrogen atom is lower than the electron energy in neutron [4]. Thus electron in hydrogen atom, while emitting and loosing energy, will never fall on proton. The Bohr's postulate does not work! For the same reason, a substance turned into plasma (i. e. non-linked electrons and protons) by heating cannot produce the original chemical element by cooling.

Sometimes they build the argument against existence of the fields consisting of equally charged based on existence of Coulomb force. But what about existence of molecules of hydrogen or other gas which consists of ionized (equally charged) atoms? Therefore, it is possible in nature. All it needs is appropriate conditions.

Let us take a simple example of electron transit in Wilson cloud chamber. In giving all of its energy to water vapors, electron is still there. It just turns into a zero energy electron, and multiple electrons with zero energy generate Dirac's electron field. Other example of such energy loss in electron is the commonly known photo effect discovered by Stoletov.

X-radiation upon electron bombardment of anode surface in a tube also argues in favor of existence of electron field. And how about well-known electron Thermionic emission from metal where electrons come from? And why the metal mass did not change given that they have left metal and took away mass with them?

In fact, for the mass of body, it is not that simple. There is an Einstein's postulate according to which rest state and state of uniform rectilinear motion are equal and they cannot be distinguished by any experiment [2]. On the other hand, according to

Lorentz transformations [3], mass of a moving body (in a uniform rectilinear way) depends on movement velocity against ether.

$$m = \frac{m_0}{\sqrt{1 - \frac{v^2}{c^2}}}$$

Where m_0 is the rest mass, v is the body velocity, c is velocity of the light in ether.

What is interesting, the recent physical experiments fit well with the of Lorentz equation. What can be the reason? We believe, there is no rectilinear and uniform movement and all interactions take place in super field (ether) rather then in vacuum. Accordingly, body moves in a medium with its own energy density and viscosity. Obviously, both parameters will resist such motion in a non-linear way. At that, the resistance force will be proportional to the square of the body velocity. Forces of adhesion are present between the moving body and the medium, so the ether layer completely stays near the body surface, as if "sticking" to it. It rubs against the next layer which is slightly comes off the body. It, in turn, is influenced by the friction from a more remote layer, and so on. The layers furthest from the body can be deemed as rest layers. Theoretical estimate of internal friction leads to **Stokes formula** [5]:

$$F_r = \frac{\rho_a S v^2}{2}$$

where F_r is the resistance force of body movement, S is the resistance area, v is the velocity of body, ρ_a is the density of the medium.

Again, the resistance force is proportional to the square of movement velocity. So while the body moves in ether (super field) it gets more real mass as if by adherence at an increased friction while increasing the area as a result of adherence of the ether layer. Besides, the super field actually compresses within the body. And that is exactly with Lorentz formula shows.

Man's falling into the water well demonstrates this. At the low height of 1 meter, i. e. at velocity V < 10 m/s, water resistance to the body moving deeper is not high, while at the height of > 100 meters the hit against the water is equal to hitting a concrete surface.

Doppler Effect speaks in favor of existence of fields in space. Einstein's claim regarding limit velocity of propagation of light (electromagnetic wave) in vacuum is well known. His approach was based on the concepts of isotopic Universe and vacuum as an empty space [2].

As the modern research of the space shows, the Universe is totally not isotropic and vacuum has a set of physical constants [5] and therefore is not emptiness. That is, vacuum is a medium so the laws of propagation of electromagnetic wave in medium applies to it.

One of the evidences of this is Doppler Effect (the red offset) in radiation spectrum of the stars moving off from us. Doppler's Effect manifests in different types of medium (water, air, and ether). This results from the fact that the medium gets denser before the moving radiator thus changing its frequency; that can be demonstrated by the Figure 4

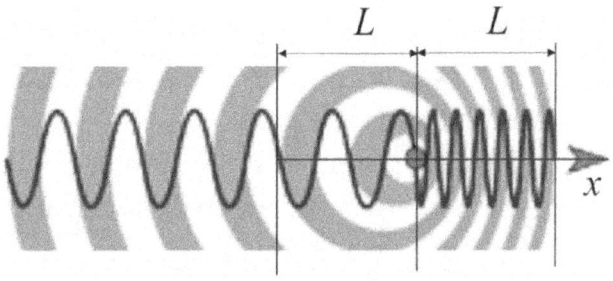

Figure 4. Doppler Effect

Frequency change of f radiation will result into change in the rate of wave propagation. This can be verified by the formulas of energy from mechanics and quantum physics:

$$E = mv^2/2; \quad (1.3)$$
$$E = hf \quad (2.3)$$

It follows:

$$v = (2hf/m)^{0.5} \quad (3.3)$$

Certainly, these equations could be applied in case when the observer is located within the axis of radiation propagation. Generally, to solve both equations jointly application of travel rates projections of the source of radiation and travel vector of the observer on the axis x is required.

Another evidence of the fact that the propagation rate of electromagnetic wave depends on the medium properties can be drawn from the experiment held by Vavilov and Cherenkov. As it shows, velocity of light propagation in a tank with distillate water is lower then the velocity of the electron brought into the tank using accelerator.

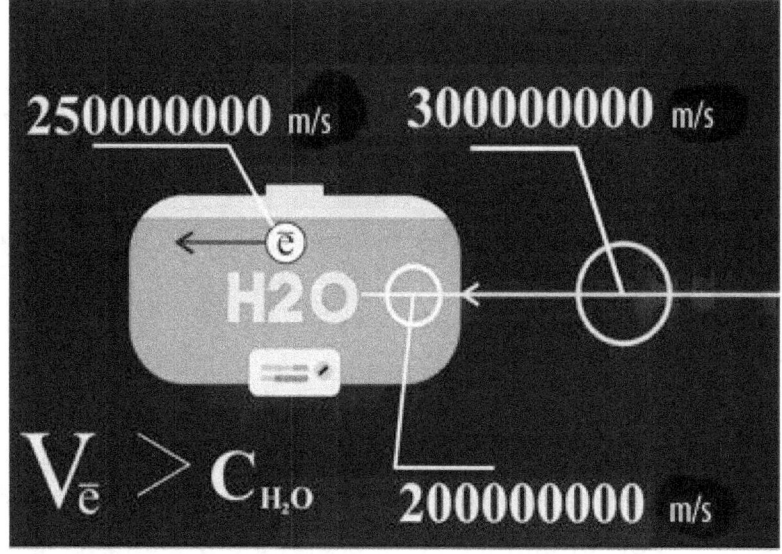

Velocity of electromagnetic wave propagation (within the visible light range) in distillate water is lower then the travel velocity of an individual electron in such water.

So the light velocity is limited only in specific medium and nothing prevents to assume that other particles exist; they would travel within the media at a rate higher then light velocity. Candidate particles could be neutrino.

This can be proven by the experiment with the use of the Swiss Large Collider when the beam of neutrino resulting from the collision of oncoming proton beams had been registered in Rome before the light radiation was registered in Bern.

This is the evidence of the fact that the velocity of the neutrino beam was higher then the light velocity in vacuum by more than 30 %. And this can not just be put down to instrument inaccuracy.

With the new experimental data, a scheme of the instrument for immediate connection at cosmic distances can be proposed. This requires a detector that registers neutrino (a water tank with optical sensors to register Cherenkov's radiation) and a neutrino beam modulator. The principle of this connection is based on the entanglement effect in neutrino. The effect manifests itself by the fact that two twin particles created in the same process are sort of connected to each other (entangled) and when one of them changes its travel parameters (by any reason) there is an instant change in the parameters of the other particle wherever it stayed. The experiment was carried out using photons (ref. page 19).

There are multiple processes in outer space that come along with the twin particle creation. The particles spread in different directions. Now if one of the particle changes its velocity as a result of placing a modulator that changes its travel parameters on its way, then the other particle changes its parameters as well wherever it would stay in the Universe. So if we take a source of twin particles (e. g. the Sun) and modulate the motion of these particles (electron, proton, neutron, photon, and neutrino) by the detector placed somewhere on Pluto or Alpha Centaurs then we can get information almost immediately rather then in 4 years.

Electric field, magnetic field, or ongoing particle beam can modulate the motion of a twin particle if we change the parameters of a damping media (liquid crystals) and so on.

Most probable that other sentient beings use this means of connection. All we have to do is place detectors on the way of cosmic particles coming from deep space and analyze the changes in their properties to decrypt information they carry.

4. Energy Exchange in Atom

Thus, based on the fact that all objects in the Universe are submerged in super field, including without limitation atoms of chemical elements, we have created the model of atom energy exchange with its environment [6].

It should be noted, that the proposed atom model suggests that the surface of rotation of the electron in spatial cone, when at some point it has a frontal orientation towards energy propagation vector in space, acts as the absorption "cross section" thus eliminating the possibility of energy quantum passing without any contact with atom. The frequency of the absorbed energy quantum is determined by the frequency of electron own vibrations.

Similar effects happen in all spatial cones that have definite parameters that, at some point, took a frontal orientation towards the energy wave propagation vector. Accordingly the geometry of spatial cones strictly influence the absorbing and radiation spectra of specific atoms of chemical elements.

As a result of the processes outlined above energy waves with a set of different frequencies that came to the atom of a specific chemical element will be absorbed and radiated depending on the geometrical parameters of its spatial cones. This results to the possible identification of each chemical element.

A stable presence of electrons in cones is reached as a result of the interaction between the energy waves propagating in space and electrons of atom. Accordingly, the source that produces photons (e. g. a star) initiates and provides for fusion of matter particles (electrons, protons, neutrons, and atoms) from photons and serves as the source of their existence because part of the wave energy that comes in from the source is spared for energy feed and spinning of electrons and spatial cones in atom.

However the issue of stable existence of atom should not be fully studied if we consider only the atom elements feeding from the energy transferred by the waves from its source. Separately we shall settle also on reasonable consumption of this energy by the elements of the atom. It is known that full energy of electron can be defined by calculation. Accordingly, it seems relevant to solve Schrödinger equation and define resonance frequencies of electron.

The place of electron within atom could be fully outlined by solving this equation. This, in turn, permits the calculation of resonance frequencies of electron radiation and absorption at their energy levels.

To define resonance frequencies of electrons at their orbits we will use well known formulas:

$$r_n = n^2\, r_1, \qquad (1.4)$$

where r_n is the radius of n orbit of electron,

n is the order number of electron orbit (inner quantum number)

$r_1 = 5.29 \cdot 10^{-11}$ m is the radius of the first Bohr orbit

$r_2 = 2^2 \cdot 5.29 \cdot 10^{-11} = 21.16 \cdot 10^{-11}$ m.

$r_3 = 3^2 \cdot 5.29 \cdot 10^{-11} = 47.61 \cdot 10^{-11}$ m.

$r_4 = 4^2\, 5.29 \cdot 10^{-11} = 84.64 \cdot 10^{-11}$ m.

$r_5 = 5^2\, 5.29 \cdot 10^{-11} = 132.25 \cdot 10^{-11}$ m.

Velocity V_e of electron rotation on the orbit r_n can be derived from the equation:

$$V_e = n \cdot h / 2\pi m_e\, r_n \qquad (2.4)$$

by substituting ratio (1.4) in the equation we have:

$$V_e = h / 2\pi m_e\, n\, r_1 \qquad (3.4)$$

where n is the number of electron orbit,

$m_e = 9.1 \cdot 10^{-31}$ [kg] is the electron mass,

$h = 6.6 \cdot 10^{-34}$ [J] is the Planck's constant.

$v_1 = 6.6 \cdot 10^{-34} / 2\pi\ 9.1 \cdot 10^{-31} \cdot 5.29 \cdot 10^{-11} = 2.183 \cdot 10^6$ [m/s] is the electron velocity at the first energy level.

$v_2 = 2 \cdot 6.6 \cdot 10^{-34} / 2 \cdot 3.14 \cdot 9.1 \cdot 10^{-31} \cdot 21.16 \cdot 10^{-11} = 1.09 \cdot 10^6$ [m/s] is the electron velocity at the second energy level.

The analysis of (3.4) shows that only n number of the orbit is a variable, while the others are the constants. It follows that the further atom is from the nucleus the less energy it has.

The Bohr quantizing rule results into the ratio:
$$n \cdot \lambda = 2\pi \cdot r_n. \qquad (4.4)$$
By substituting for $r_n = n^2 r_1$, we will have:
$$\lambda = 2\pi \cdot r_1 \cdot n \qquad (5.4)$$
When we know that $f = c/\lambda$ we can write down the expression to determine resonance frequencies of electrons at all known orbits.
$$f_n = c/2\pi \cdot r_1 \cdot n \qquad (6.4)$$

For hydrogen atom, $f_1 = 3 \cdot 10^8/2\pi \cdot 5{,}29 \cdot 10^{-11} = 3 \cdot 10^8/3.32212 \cdot 10^{-10} = 0.91 \cdot 10^{18}$ Hz.

This fits with the experimental data.

We shall try to calculate full energy of electron. We will examine as an example Li (lithium) atom interaction with its environment.

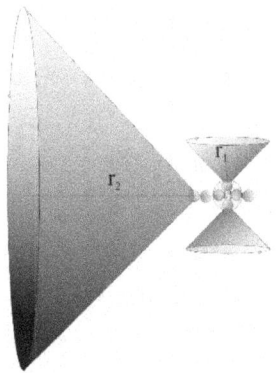

Figure 5. Lithium

For lithium, resonance frequencies are:

$f_1 = 0.91 \cdot 10^{18}$ Hz,

$f_2 = 3 \cdot 10^8 / 2\pi \cdot 5.29 \cdot 10^{-11} \cdot 2 = 0.455 \cdot 10^{18}$ Hz

Because electrons move not in a vacuum but in ether media, friction occurs; atom cone will absorb energy to compensate the friction by cutting the whole range of coming frequencies down to the frequencies it needs, i. e. resonance frequencies.

From the Planck's equation $E = h \cdot v$ we can write down:

$$E_1 = h \cdot v_1; \; E_2 = h \cdot v_2 \rightarrow \Delta E = h(v_1 - v_2) \rightarrow$$
$$v_{rad.} = (v_1 - v_2) \qquad (7.4)$$

This means that radiation frequency in atom excitement equals to the difference of frequencies in according cones. For lithium $_3Li^7$:

$v_{rad} = (0.91 - 0.455) \cdot 10^{18} = 0.455 \cdot 10^{18}$ Hz.

The obtained value matches with the experimental data.

As it is known from the physics course, "moving" waves have a higher attenuation constant. Standing waves, in contrast, have significantly less attenuation due to the lack of interaction with the environment and their property of being "closed in themselves".

As we mentioned above, electron creates the surface of circular still waves of definite frequency when it travels along its orbit in spatial cone at defined velocity. In addition, as it was mentioned above, such surfaces serve as frequency filters that distribute energy waves. The wave energy with the frequency equal to the filter frequency will be absorbed by electron. Regardless of how much energy gets onto this filter, its frequency properties will keep the same. It shall be noted that electron moves along its orbit in sinusoidal way with a stable frequency but with a variable magnitude. This is schematically shown in Figure 6.

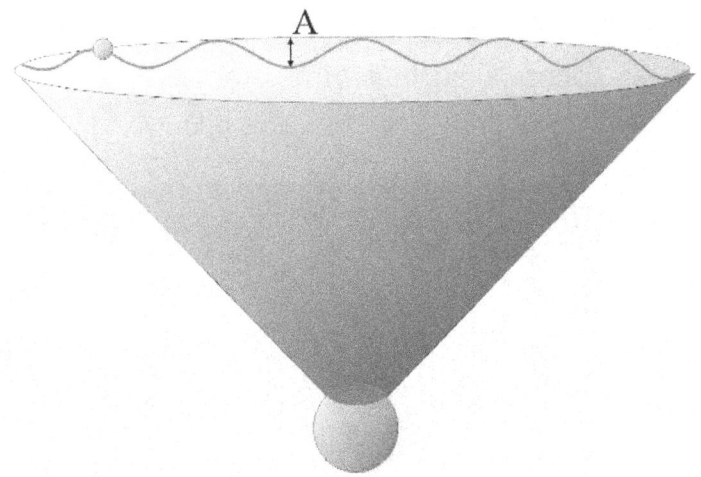

Figure 6. Scheme of spatial-cone motion of electron along its orbit

Where A is amplitude – invar.

Accordingly, the energy wave, which has approached to the cone, can cause an increase in its frequency amplitude without any effect on its oscillation frequency.

The atom form based on the proposed model supported by the stability of its existence from the energy perspective, the solution of Schrödinger equation as provided above, and the definition of resonance frequencies of electron, has every right to be considered for every solid scientific discussion. The more so as the atom level photographic images of such commonly known chemical element as Aurum made by Japanese scientists do not prove the spherical surface of its atoms.

a)

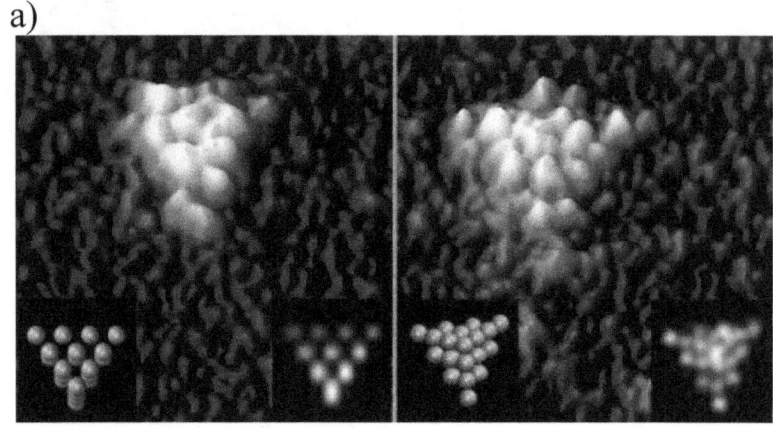

b)

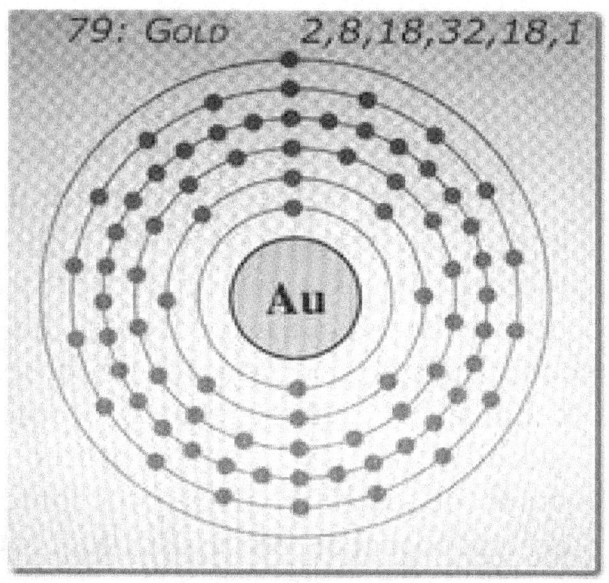

Figure 7. Photo of Au atoms group (a) and scheme of atom form (b)

Comparative analysis of the current atom model and the model proposed herein calls for continuation. While the laws of energy wave propagation within the proposed spatial/cone atomic model [6] work completely in alignment with the physics laws, this is impossible to claim the same in respect of wave interaction with atom as per Bohr's model. So, we will analyze how sinusoid wave interacts with the spatial/cone model of atom. After the straight wave comes into the "absorption cross section" it will pass through a spatial cone and then will reflect at the same frequency but with a phase shift by π. Accordingly, there is everything necessary for wave energy to effectively propagate; this is much the same as radio waves propagation using the very reasonable antenna size.

In contrast to the Bohr's model, the spatial/cone model of atom "peacefully" fits with energy exchange with environment. In getting energy from outside, this atom absorbs it with an increase with its inner energy and causing an increase of space deformation. Further energy consumption by radiation and own rotation causes the change in space deformation. This is followed by energy absorption and the process starts all over again.

Accordingly, atom "breathes" only because of the energy received and, generally speaking, "lives" as an autonomous matter spatial entity completely dependent from what was source of itself and is the source of its "food".

5. Gravity

As we have mentioned above, the photon field is accelerating with the vector directed to the Earth center as a result of the fusion. It follows that the elementary particles composing the material objects are interacting with the accelerating field and the gravitational force evolves, and its vector matches the vector of the field. Obviously, the concentration of fields of which ether consists can be very much different depending on location in the Universe. For example, γ field will be the main energy component near the Sun and it will define gravity to the most extent. According to the above mentioned, we can study gravity as an interaction of the objects with the medium (ether) that itself takes part in energy and matter conversions.

The concept of the body mass is a reasonable place to start the study of gravity. In mechanics, the body mass is the quantity of the neutrons, protons, and electrons in this body and it manifests itself in mechanical interactions.

The inertial mass of the body can be also determined by the number of the neutrons, protons, and electrons in this body and it manifests itself only in attempting to change the body travel velocity in relation to the]on field; i. e. inertia is an interaction of the body moving with acceleration in relation to the surrounding field. The inertial mass of the material body manifests itself regardless of the presence of other bodies nearby.

Gravitational mass is a set of neutrons, protons, and electrons in the specified body quantity and it manifests itself in the interaction with the moving field respective to the presence of other material bodies in it.

Everyone knows from their schoolroom days, that the most visual way to demonstrate the gravity forces and free-fall acceleration is the Newton's tube; this device is a 1-meter glass tube which has one end welded up and the other end equipped with a valve. To perform the demonstration experiment, three different objects are placed into the tube, e. g. a shot, a cork, and a feather. Then the tube is turned upside down very quickly. All three bodies will drop to the bottom of the tube but at various times — first, the shot, then the cork, and finally the feather.

This happens if air is present in the tube. But as we evacuate air from the tube with a vacuum pump and turn it upside down, we can watch how all three bodies reach the bottom of the tube at the same time. Air evacuation from the tube has resulted in the elimination of the resistivity of air for all three bodies, regardless of their weight and volume, so that they were affected only by the gravity force with the free-fall acceleration similar for all bodies.

The examination of the above experiment gives us, additionally to the described above, the note that this experiment is also the most visual way to demonstrate the transportation of the ether medium at the Earth surface. Air evacuation from the tube did not affect the transportation of the ether medium within the medium and the interaction of the ether flux with the electrons, protons, and neutrons of the free-falling bodies.

Accordingly, air evacuation from the tube has eliminated the slowing-down effect of the air with the various force on the different objects; this has enabled the creation of the most favorable conditions to secure the purity of the experiment conducted upon determination of the effects of the ether flux on the elementary particles that combined in various ways in building of the specific material bodies, regardless of what kind of body it was — the shot, the cork, or the feather. In this case, they all move similarly in the general ether flux regardless of their shape and sizes, just like a log and a match would move similarly in a general river flow.

So, we assume that, among other matters, the experiment that uses the Newton's tube with the evacuated air can qualitatively illustrate not only the movement of the ether medium towards the center of the Earth (planet) but also allows us to make quantitative calculations of its acceleration, which is approximately equal to 9.81 m/sec^2.

The free-fall acceleration can be approximately calculated (in m/ sec^2) by the empirical formula:

$g = 9.780327(1 + 0.0053024 \, \text{Sin}^2\varphi - 0.0000058 \, \text{Sin}^2\varphi) - 3.086 \cdot 10^{-6} \cdot h$, where φ – the latitude of the location under consideration, h – the height above seal level in meters

The obtained value matches only approximately the free-fall acceleration at this location. The quantitative values of the acceleration at the various heights h above sea level are shown in Table 1.

Table 1

h, km	0	5	10	15	20	50	100
g, m/s²	9.806	9.791	9.775	9.760	9.745	9.654	9.505

h, km	500	1000	10^4	$5 \cdot 10^4$	$4 \cdot 10^5$
g, m/s²	8.45	7.36	1.5	0.125	0.002

The values in the table give evidence that the free-fall acceleration value is not constant but is increasing as it approaches the center of the Earth. This fact also confirms the relevance of the above calculation and the proposed assumption regarding the reasoning behind the fusion of the electrons from the ether field in the Earth interior.

From these, we can easily understand what the gravity force represents. The force is known to occur when the "mass" of the body and the acceleration are present. In this way, any body, which has a mass and is located near a planet (including the Earth), will be, at the same time, automatically present in the accelerating medium directed towards the center of the planet.

By multiplying the mass of the material body under consideration by the acceleration of the ether medium, which we customary call the free-fall acceleration, we will obtain the force at which the body is held against the planet by the ether flux. This force is also known as the gravity force.

The fullness and integrity of the scientific approach that we propose and that is based on the analysis of the effects, processes, and interactions in the ether medium shall be complete with the reasoning for the attraction of two bodies at the Earth surface.

Using the Newton's law of universal gravitation, let us set down the equation to calculate the gravitational force that affects the material body:

$$F = \gamma \cdot m_{Earth} \cdot m_2/R^2 \ (1)$$

where: $m_{Earth} = 5,9742 \cdot 10^{24} \ kg$ – the Earth mass
$\gamma = 6.67 \cdot 10^{-11} \ m^3 kg \cdot f^2$ – gravitational constant
m_2 – the mass of the body.
$R = 6,371,302 \ m$ – the mean Earth radius

At the other hand, the two bodies are attracted to each other, probably, according to the same law, see Figure 7.

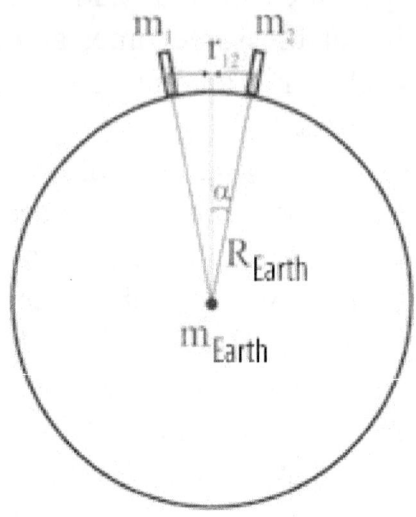

Figure 7. Attraction of two bodies

Let us assume $m_1 = m_2 = 100$ kg, $r = 1$ m.
$$F_{Earth1} = F = \gamma \cdot m_{Earth} \cdot m_1/R^2 = F_{Earth2}$$
$$F_{Earth1} = 6.67 \cdot 10^{-11} \cdot 5.9742 \cdot 10^{24} \cdot 100/(6371302)^2 =$$
$$= 39.847914 \; 10^{15}/40593489175204 = 986 \text{ N.}$$
$$F_{12} = \gamma \cdot m_1 \cdot m_2/r^2$$
$$F_{12H} = 6.67 \cdot 10^{-11} \cdot 100 \cdot 100/1^2 = 6.67 \cdot 10^{-7} \text{ N.}$$
At the other hand, the force F_{12} can be determined as a pressurizing force from the two vectors of the force of the ether flux along these bodies towards the Earth.

$$F_{12pr} = F_{Earth1} \cdot \sin\alpha$$

$$\sin\alpha = (r_{12}/2)/R$$

$\sin\alpha = 0.5/6371302 = 7.84768953 \cdot 10^{-8}$

$F_{12pr} = 986 \cdot 7.84768953 \cdot 10^{-8} = 7.74 \cdot 10^{-5}$ N.

Accordingly, the pressurizing force F_{12pr} from the two vectors of the ether flux force of these bodies towards the Earth is a hundred times higher than the force of attraction between the two bodies F_{12N}, calculated as per Newton's law.

The example soundly shows that the nature of the initiation of the attraction between bodies in the field of the gravitational object can be determined quite logically by the theoretical prerequisites.

6. Particles from the Field

As was mentioned above fields consist of particles that have the same parameters. The filed particles have minimum possible energy (particle energy tends to zero). Therefore, the filed consisting of such particles (dark matter) interacts infinitely little with material bodies. However, it does not mean that it could be ignored. The reason is that the field creates the matter as a result of fusion. And material particles mass is defined during the fusion. At this, self-energy density of the created particle is inversely proportional to its mass. This means that the densest fields that consist of smaller particles contain more energy in volume unit (dark matter). This can be proven by the large difference between the energies produced in hydrogen oxidation (this is combination of atoms H and O, i. e. fusion): $E_{H_2O} = 1.20 \cdot 10^8$ J/kg. As well as in hydrogen bomb explosion (fusion of He from D) $E_{He} = 10^{15}$ J/kg. This means that energy density of thermonuclear explosion is 10^7 higher then the density energy of hydrogen burning Energy density of electron fusion from photons $E_e = 3.38 \cdot 10^{32}$ J/kg (theoretical) is $3 \cdot 10^{17}$ higher then the energy density of thermonuclear bomb explosion. The figures prove our assumption. And if we take smaller objects, such as photons, their inner energy is even higher. Accordingly, we are surrounded by the ocean of infinite energy. This ocean provides for constant creation (as a result of fusion) new particles. How some particles are created is shown in Figure 8.

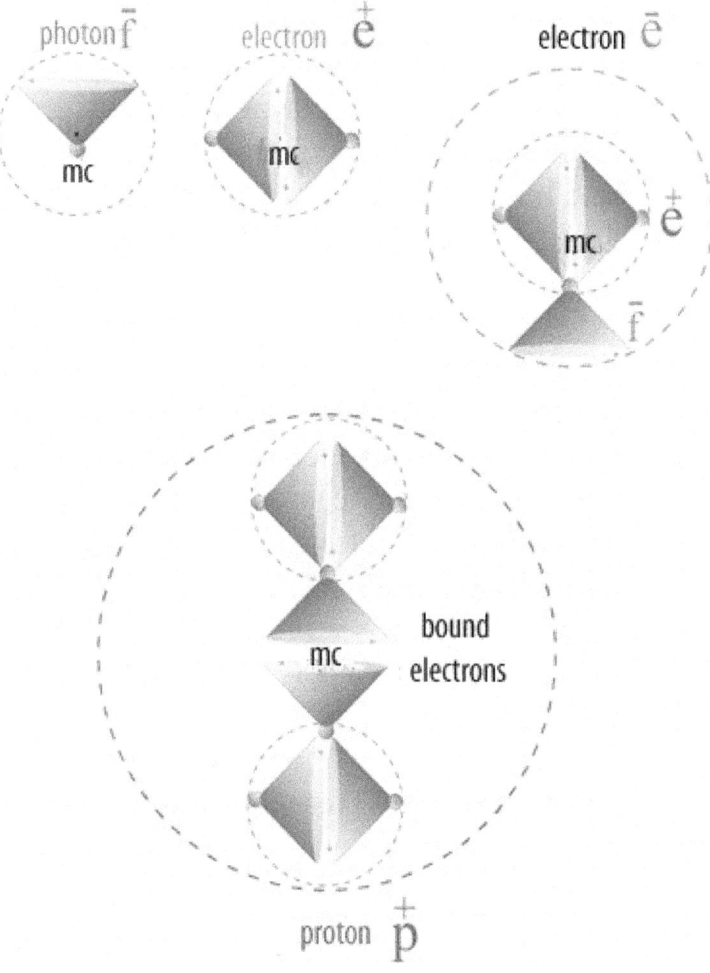

Figure 8. The diagram of particle creation

43

It should be noted, that mass center (mc) of photon is located near the nuclei because the nuclei mass is 1836 higher than the satellite mass. This means that the negatively charged area will shield positive charge of the nuclei upon photon rotation around the mass center.

When two photons are paired (in analogy with chemical bond) they synthesize positron. Positively charge of nucleus will shield negative charge of the satellites upon the photon pair rotation around the common mass center. When positron takes another photon, electron is created. Its negative charge forms externally because of the rotation of photon cone around common mass center of the system (electron). When it takes another electron, the system of bound electrons is created (proton). Its positive charge is caused by the fact that negatively charged parts of electron are shielded by the positive charges in nucleus.

For the purposes of information and convenience of calculations, most appropriate may be proton which consists of two electrons; for this reason we will calculate the velocity of electromagnetic wave propagation in proton using the expression of relativistic mass:

$$m_p = \frac{2m_e}{\sqrt{1 - \frac{v^2}{c^2}}} \qquad (1.6)$$

After having the transformation done, we will have:

$$v_e = \sqrt{c^2 \left(1 - \left[\frac{2m_e}{m_p}\right]^2\right)} \qquad (2.6)$$

Using the fact that $m_p/m_e = 1836$, we will have:

v_e =c· 0.99986=29979245·0.99986= 29975163.3 [m·s-1]

Then we will calculate electromagnetic wave frequency (circular wave of electrons) in proton:

$$v_p = v_e / 2\pi r_p, [c^{-1}] \qquad (3.6)$$

where v_e is velocity of wave propagation in proton, [m·s^{-1}]

r_p is proton radius = $8.751 \cdot 10^{-16}$ [m]

v_p = $2.9975163 \cdot 10^8$ / 2π · $8.751 \cdot 10^{-16}$ = $2.9975163 \cdot 10^8$ / $5.4956 \cdot 10^{-15}$ = $0.5454 \cdot 10^{23}$ [c^{-1}]

We will calculate the length of electromagnetic wave in proton λ, [m]
$$\lambda = c' / v_p, \qquad (4.6)$$

because $c' = v_e$

45

$$\lambda_p = v_e / v_p = 2.9975163 \cdot 10^8 / 0.5454 \cdot 10^{23} = 5.495996 \cdot 10^{-15} \text{ [m]}$$

The obtained value is accurate to the fourth decimal and matches the proton equator, which means that electromagnetic wave in proton travels along the spherical surface of proton.

This fact gives evidence in favor of conformity of the proposed model of particles based on the example of proton with the estimation data. Additionally, the hypothesis proposed herein can be proven by the results of the experiment carried out in 2015 using the Large Hadron Collider. Two beams of protons had broken-up upon collision:

$$p^+ + p^+ = 2e^- + 2e^+ + 2\gamma,$$

This means that proton consists of electron, positron, and γ quantum (photon). See Figure 8.
This is well in line with the Standard Model in Particle Physics.

The only thing left is to explain how the linkage modeled on chemical bond is formed. We will see Figure 9.

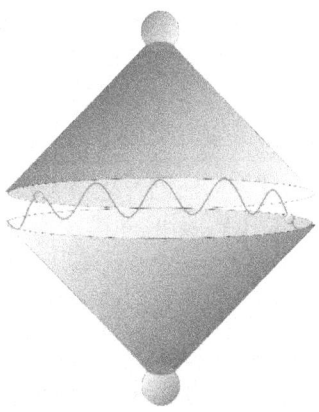

Figure 9. How the linkage in analogy with chemical
bond is formed

"Electrons" get in turns into the zones of
opposite nuclei when they travel by a sinusoidal
trajectory. This means that the same "electron"
always belonging to one nuclei and belonging to
another nuclei another moment. Accordingly, a new
stable dynamic system consisting of two valence
electrons and two atom nucleus is formed.

7. Energy Transfer as the Medium Deformation

As the electromagnetic wave is a sum of the conversion of electric deformation of the magnetic deformation of the same level (or of the same field) and then the magnetic wave converts back to the electromagnetic wave, then, in a simplified way, it represents an oscillatory circuit that moves over the space from the transmitter to the receiver.

The conclusion is that the flux of the superfield deforms due to the fusion resulting in creating of the electromagnetic waves in the electron field. The converged electromagnetic waves form the magnetic poles while the expanding waves are the own frequency of the planet. (The Earth, for example, has the own radiation frequency that is equal to 7.2 Hz.)

It should be noted that the wave propagation is possible only in an elastic medium where alternate compression and refraction of the medium followed by the simultaneous acceleration and slowing-down of the particles that build it.

The gravity of spatial objects is determined by the nature of the fusion that takes place in these objects. The gravity process is the actual movement of the ether to the fusion center at acceleration. It is followed by the electromagnetic radiation of a strongly defined frequency. (See Figure 10.)

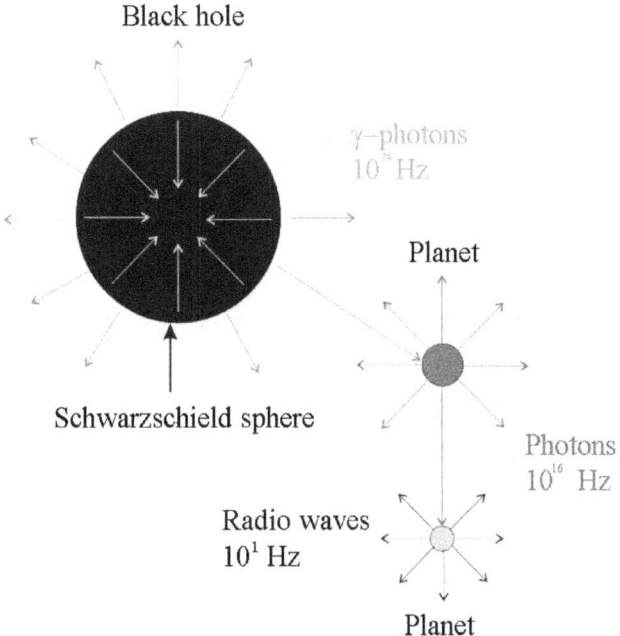

Figure 10. The gravity and radiation frequencies of various spatial objects

The gravity develops as an interaction of the objects within the ether medium that itself takes part in energy and matter conversions. The gravity process is the interaction of the object with the accelerated flux of the superfield.

We have studied the issues of gravity, electromagnetic wave, internuclear interactions that were demonstrated by the example of the formation of chemical elements [6], as well as of the interaction of spatial objects. The issue of charge still remains uncovered. In our belief, the elementary charge is a lowest-stable aggregation of the ether elements with a higher density (minus) or a relatively lower density (plus) that is involved in a vorticose movement — right or left respectively; it simultaneously forms a closed energy system.

When T-shaped particle (Minkovsky) moves in ether medium it takes shape of an egg. It has a denser front part and less dense rear part. This is caused subject to differential pressure evolving upon its flowing by field, ref. Figure 11.

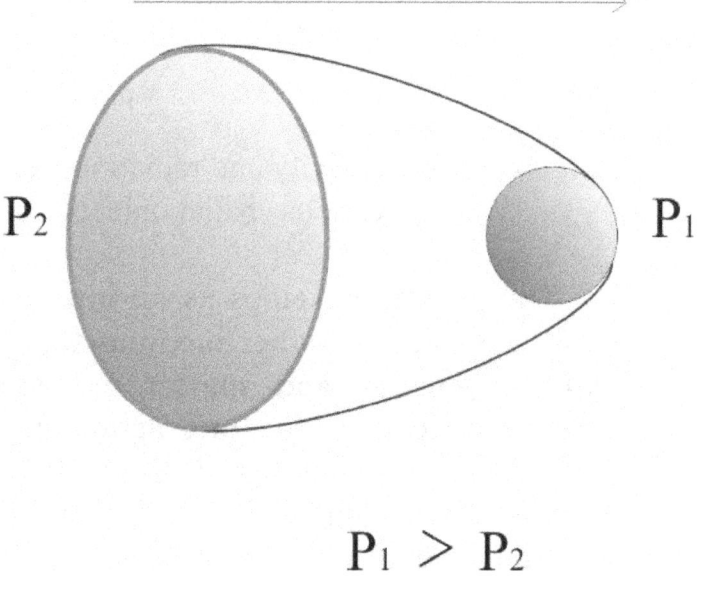

$$P_1 > P_2$$

Figure 11. Particles motion in ether

8. Conclusions

1.   **Ether is a superfield that includes all fields and particles.**
2.   **Particles of any field have a complex structure.**
3.   **All interactions are created by the deformation of ether.**
4.   **Interaction laws in different fields can be described by the same equations. The field strength in all cases decreases proportionally to the square of the distance between the objects, Coulomb forces or gravity forces in accordance with Newton.**
5.   **The Universe itself has originated from the fusion, which, similarly to black holes, stars, and planets, has initially formed at an infinitesimal point. Thus, we have the expanding Universe. There was no packed Universe in an infinitesimal quantity but rather there was an initiation of the synthesis of matter from the field.**

And there the question of what is time is still remaining. Time is a process of changing energy state of the object. **It depends on the acceleration at which an object or a field comes through it causing a change in the energy density of the object.** The higher is the density, the slower time passes. Accordingly, time is the function of the energy density of the object (subject). **A good example of this is the neutron — it can stay for billion years in atomic nucleus while it stands only for 600 seconds in the free state.** Any clock, whether electronic or mechanical, slows down in mines but it will be running ahead of time in the orbital space station, that is an established fact.

9.   Analysis of Ether Medium

The advances of the contemporary science with regard to the registration of gravity waves give a proof of the fact that such waves, equally to any other waves, can propagate the energy over astronomically large distances in an effective way, and, moreover, to exist during extremely long intervals. However, the report materials regarding this issue say nothing about the medium which made the propagation of the waves possible. Accordingly, in the light of the circumstances of propagation of the registered waves when their energy has not been dissipated for a long time (has not changed to the internal energy), and even from the common point of view of contemporary scientific approach, all the above said shall occur in an absolutely elastic medium. That is why it should be noted that if a continuous medium has elastic properties then the motion of the points at one location of the medium (in the source) well lead to the propagation of this motion at a certain velocity in the shape of an elastic wave. In addition, if the medium has only its cubical elasticity, then only the longitudinal waves (liquid, gas) can propagate; and if it has also shape elasticity, then the transverse (shear) waves are also possible. This just indicates that the medium of ether has all the above-specified properties because it can propagate not only longitudinal waves but also transverse (shear) waves. Cumulatively, the first and the last create electromagnetic waves. In addition, both the longitudinal and the transverse waves can exist independently of one another.

It should be noted separately, that the propagation of any waves without a medium (in the vacuum) is not possible. In accordance to the above said, the study of the medium where the propagation of the waves of all kinds takes place shall be considered as a separate matter.

First of all, it should be noted, that any registration instrument has a strongly limited measuring range. While the currently available measuring tools allow registering the neutrons, protons, and electrons individually by the specific particle characters, the photons, for example, the photons in a thinner level, cannot be registered by the current tools as the individual particles; this does not mean, though, that they do not exist. For example, if we had made a hot-water injection to the water, then the volume and mass of the ocean would change, the microwaves would arise, the temperature in ocean would change, but no tool can ever register all this. Accordingly, all the energy/mass exchange processes that continuously take place and the quantity of which is countless use a thinner level as an endless source of energy for one part, and for the second part as utilizer of any excess energy. The above said gives us the ground to conclude that energy is a sum of the particles in the thin-layer ether that are involved in energy/mass exchange processes and remain in an unstable state. This leads us to a conclusion that an individual Planck's constant shall be available for each ether level.

Presently, the capabilities are available that ensure the registration of the propagating waves of electromagnetic radiation and the determination of light pressure to which the reflection and absorption bodies, particles, as well as independent molecules and atoms, are exposed. In addition, while the wave properties of light are regulated by the wave laws that involve elastic media, it currently is not possible to register the corpuscular properties of light for each individual component of the particle due to a lack of capability to register these individual particles at the moment. Meanwhile, the sum pressure of the particle flux had been registered back in 1899 by P. N. Lebedev.

The above material gives a proof that, firstly, the propagation of the registered gravity waves shall involve some medium. However, the issue of whether it is correct to regard these waves as the gravity waves can naturally arise. In fact, there can be no doubt in the registered near collision of two black holes followed by their merging. The above merging shall very likely occur at least in two levels followed by the changing from one to another. We will analyze the results of LIGO experiments to discover gravity waves, ref. Figure 12.

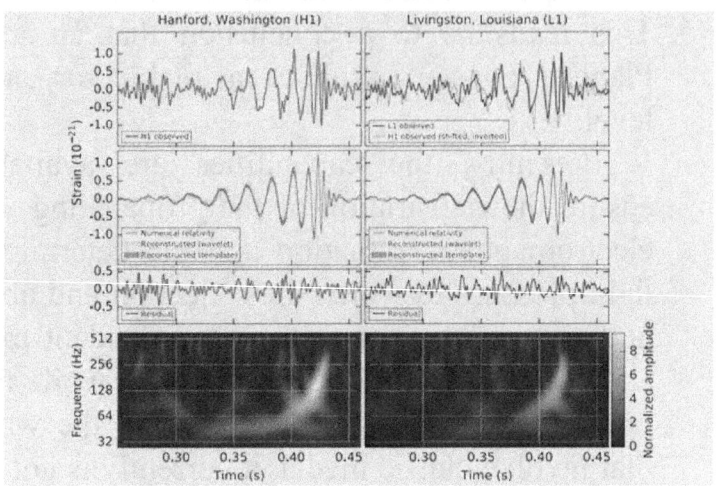

Figure 12. Results of LIGO experiment

As oscillograph recordings show, maximum energy density of the registered waves is applicable to the frequency range from 100 to 200 Hz. This corresponds to the wave length range of 1500–3000 kilometers. Obviously, to reliably register such waves, the length of the antenna shall not be less than one fourth of the wave length, i. e. > 400 kilometers while the distance between interferometers was 4 kilometers. This caused the lower certainty of identification of the obtained result with gravity waves. We will study gravity waves in more detail in our next book.

10. P.S.

Having considered all materials presented herein, we are led to the conclusion that the Universe is a complex self-developing system while the man is only a part of the system, but the part will never be more complex than the whole. Accordingly, the human mind is only a part of the Universe mind!

Reference List

1.   Einstein, A. Sobraniye nauchnyky trudov v chetyrekh tomakh [A Collection of Scholarly Works in Four Volumes] (In Russian), M.: Nauka, 1965–1967

2. Groh V. (2018) *Developmental Universality and Unity of the Universe.* 62p.

3. Groo, V. Ya., Sheshukov, V. V., Yegorov, G. I. Novaya teoriya stroeniya yader khimicheskikh elementov [The New Theory of Atomic Structure of Chemical Elements] (In Russian) https://www.youtube.com/watch?v=NZAJXfgSr1I

4.   Ivanov, I. S. Elementy bolshoy nauki / Neitrony raspadayutsya i s izlucheniyem fotonov [The Elements of Big Science / Neutrons Break Down also with the Photon Emission] (In Russian) http://elementy.ru/news/165038

5. Matveyev, A. N.(1989) *Atomnaya fizika. Uchebnoye posobiye dlya studentov vuzov.* [Atomic Physics. Tutorials for Students of Higher Education] (In Russian.). M.: Vysshaya shkola. 439 p.

6. Yavorskiy, B. M (1991). *Spravochnik po fizike* [The Physics Handbook] (In Russian) / Yavorskiy, B. M. and Detlaf, A. A. M.: Nauka,. 942 p.